MW00615664

Construction Zone

Cranes

by Rebecca Pettiford

Ideas for Parents and Teachers

Bullfrog Books let children practice reading informational text at the earliest reading levels. Repetition, familiar words, and photo labels support early readers.

Before Reading

- Discuss the cover photo. What does it tell them?
- Look at the picture glossary together. Read and discuss the words.

Read the Book

- "Walk" through the book and look at the photos. Let the child ask questions. Point out the photo labels.
- Read the book to the child, or have him or her read independently.

After Reading

- Prompt the child to think more. Ask: Cranes are big machines. They lift and move heavy things. Do you know other big machines that do this?

Bullfrog Books are published by Jump!
5357 Penn Avenue South
Minneapolis, MN 55419
www.jumplibrary.com

Library of Congress Cataloging-in-Publication Data

Names: Pettiford, Rebecca, author.
Title: Cranes / Rebecca Pettiford.
Description: Minneapolis, MN: Jump!, Inc., [2023]
Series: Construction zone | Includes index.
Audience: Ages 5–8.
Identifiers: LCCN 2021053221 (print)
LCCN 2021053222 (ebook)
ISBN 9781636908496 (hardcover)
ISBN 9781636908502 (paperback)
ISBN 9781636908519 (ebook)
Subjects: LCSH: Cranes, derricks, etc.—Juvenile literature. | Hoisting machinery—Juvenile literature.
Classification: LCC TJ1363 .P45 2022 (print)
LCC TJ1363 (ebook) | DDC 621.8/73—dc23/eng/20211209
LC record available at https://lccn.loc.gov/2021053221
LC ebook record available at https://lccn.loc.gov/2021053222

Editor: Jenna Gleisner
Designer: Michelle Sonnek
Content Consultant: Ryan Bauer

Photo Credits: Bjorn Heller/Shutterstock, cover; Lalocracio/iStock, 1; Umlaut2013/Dreamstime, 3; MODSTEEL/Shutterstock, 4; jantsarik/Shutterstock, 5; aapsky/Shutterstock, 6–7; nano/iStock, 8–9; ZoranOrcik/Shutterstock, 10–11, 17, 23tl, 23bm; Lobachad/Shutterstock, 12, 23bl; Nick Rostov/Shutterstock, 13; thaloengsak/Shutterstock, 14–15, 23tm; Fahroni/Shutterstock, 16, 23tr, 23br; Budimir Jevtic/Shutterstock, 18–19; Roman023_photography/Shutterstock, 20–21; Vereshchagin Dmitry/Shutterstock, 22; Sanit Fuangnakhon/Shutterstock, 24.

Printed in the United States of America at Corporate Graphics in North Mankato, Minnesota.

Table of Contents

Lift It Up

Cranes are big machines.
They lift heavy things.

Some are tall.

They stay in one place.

Others have wheels.

They move!

wheel

Cranes help us build.

outrigger

Outriggers go out.

They keep the truck still.

Pat moves the joysticks.

They move the crane's parts.

The boom goes out.
It is like a big arm.

boom

It has pulleys.
Hooks are on the ends.

A hook picks up a beam.
beam

Look!

The boom lifts the beam.

It is up high!

It sets the beam down.

It will go back for more.

Cranes do big jobs!

Parts of a Crane

What are the parts of a crane? Take a look!

Picture Glossary

beam
A long, thick piece that helps support a building.

boom
The long arm on a crane that moves large or heavy things.

hooks
Curved pieces of metal used for hanging things.

joysticks
Levers that are used to control a crane.

outriggers
Legs on a crane that keep it in place.

pulleys
Wheels around which cables can turn to lift heavy loads.

Index

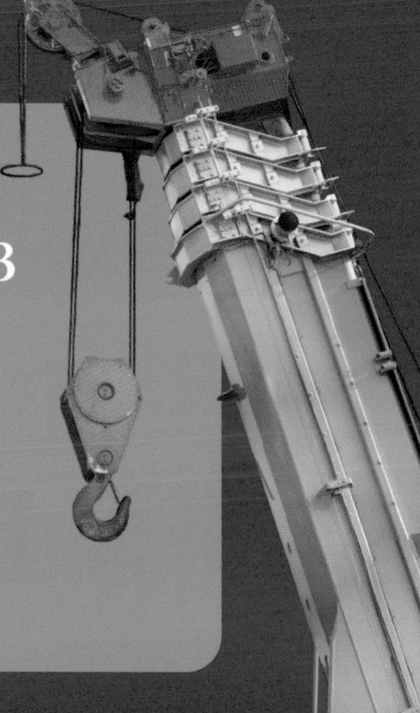

To Learn More

Finding more information is as easy as 1, 2, 3.

1. Go to www.factsurfer.com
2. Enter "cranes" into the search box.
3. Choose your book to see a list of websites.